THE ISTHMIAN ROUTES.

A BRIEF DESCRIPTION

OF EACH PROJECTED ROUTE, AND OF THOSE NOW EXISTING, SHOWING THE CAPACITY OF THEIR HARBORS, THE COMPARATIVE ADVANTAGES OF EACH, AND THE DISTANCE BY EACH FROM

NEW YORK TO SAN FRANCISCO,

FROM THE BEST SOURCES OF INFORMATION, AND FROM PERSONAL OBSERVATION AND SURVEY OVER EACH, IN THE YEARS

FROM

1856 to 1861.

By CAPTAIN SEYMOUR.

NEW YORK:

HALL, CLAYTON & MEDOLE, PRINTERS, 46 PINE STREET.

1863.

THE ISTHMIAN ROUTES.

A BRIEF DESCRIPTION OF EACH OF THE PROJECTED ROUTES, AND OF THOSE EXISTING—SHOWING THE CAPACITY OF THEIR HARBORS, THE COMPARATIVE ADVANTAGES OF EACH, AND THE DISTANCE BY EACH FROM NEW YORK TO SAN FRANCISCO.

These Routes, commencing with the most Northern and proceeding Southward in the order in which they occur, are as follows:

1. *The Tehuantepec, through Mexico.*
2. *The Honduras,* " *Honduras.*
3. *The Nicaragua,* " *Nicaragua.*
4. *The Chiriqui,* " *Isthmus of Chiriqui.*
5. *The Panama,* " *Isthmus of Panama.*

1. THE TEHUANTEPEC ROUTE.

The Atlantic terminus of this route is at the mouth of the Coatzacoalcos River, which empties into the Gulf of Mexico, in Lat. 18° 8′ N., Long. 94° 31′ N. from Greenwich, and distant from Vera Cruz about 120 miles. The Coatzacoalcos has a bar at its mouth upon which there is but 11 feet water at low, and 13 feet at the highest tides. The coast at this point forms an extended indentation or "bight," by stretching eastward. The current across the mouth of the river is strong, setting from East to West. The prevailing winds are from the North, and during Northern gales the land becomes a dead lee shore, from which sailing vessels cannot escape except by venturing the passage of the bar, which is almost impossible, as it shifts with the storms, and there is no certainty of a passage during their continuance.

During heavy "Northers," the surf on the coast is so high that pilots are unable to board vessels, whatever their distress or danger may be. Vessels of the lightest draft of water are frequently wrecked or disabled in crossing the bar, by grounding between the swells of the sea. The steamer "Quaker City," of 11 feet draft, struck twice on the bar in 1858, on her second trip with the California mails, and, after this, was forced to lie outside and deliver them by the aid of a small vessel, the "Jasper;" and, on one occasion, she had to return to New Orleans without effecting a landing.

The coast is such that every gale down the valley of the Mississippi becomes a "Norther" near the mouth of the Coatzacoalcos, and prevents, practically, all crossing of its bar, or communication with "Minititlan," the port of entry, which is 24 miles above the bar.

From Minititlan to Suchil, the head of navigation, is 90 miles by the river; and from Suchil to Ventosa, the terminus of the route on the Pacific, 117 miles; making the whole distance of the route across, 231 miles.

The river Coatzacoalcos is shallow, with a shifting bottom, and to make the navigation of the 90 miles at all successful will require the removal, not only of the numerous snags, but the use of boats drawing less than two feet of water. On the land route, the highest elevation above the Pacific is 810 feet—at La Chivela Pass. The grades are 71 feet to the mile. Six bridges are required, viz.: over the rivers Jaltepec, Jumuapa, Sarabia, Malatengo, Los Perros and Tehuantepec, with spans of from 250 feet to 600 feet.

On the Pacific slope, the mean temperature is 81° Fahrenheit, with annual variations from 76° to 84°. The climate from the Pacific shore to the base of the Cordilleras is, in general, warm, dry, and comparatively healthy. On the Northern or Atlantic slope, the temperature is more variable; the thermometer frequently ranging at 92° at seven o'clock in the morning. The climate is damp and unhealthy; intermittent fevers are prevalent, and the "black vomit" constantly exists. The productions on the Atlantic, or Northern slope, are the india-rubber tree, the ixtle plant, Indian corn, cochineal, sweet potatoes, and most of the tropical fruits. On the Pacific, or Southern slope, the most valuable products are Brazil wood, campeachy, palm, of various species, cocoa, lignum-vitæ, mahogany, cedar, rosewood, and the mosquito, producing gum-arabic.

The Pacific, or Southern terminus of this route, is at Ventosa, in Lat. 16° 21′ N., Long. 95° 2′ W. from Greenwich. This is a roadstead open to the sea from the S. S. W. to E. N. E.; depth of water from 20 to 59 feet; anchorage good, bottom being of sand and mud; rise and fall of tide, 4 feet, 8 inches.

The prevailing winds are S. W. and N., from the month of November to August; and during the months of August, September and October, there are hurricanes from the S. and S. E., during which no ship can be safe at anchor.

The currents set up the shore, and out S. S. W., about one mile per hour; the sea at the landing is almost always heavy, and frequently it is impossible to land with the best surf-boats. In the years 1858 and 1859, the steamer Oregon, after landing the mails and passengers, left for a more sheltered point further up the coast to await the return mail. A permanent breakwater has been proposed, and there is no doubt this could be constructed at Ventosa, or at Salina Cruz; but the cost would be enormous—estimated at twenty-five millions of dollars, and would involve years in construction, as well as the most complicated and difficult problems of hydrographical engineering.

Ventosa was made a port of entry in 1857, but up to July,

1859, there had only been nine arrivals of foreign sailing vessels, of which two were wrecked, and three parted their cables and were drawn to sea.

THE HONDURAS ROUTE.

The Atlantic or Northern terminus of this route is Porto Caballos, in Lat. 15° 49′ N., Long. 87° 57′ W. from Greenwich. This port is about eight miles in area, but is unsheltered, open, and exposed to the sea from the Westward. Depth of water from 4 to 12 fathoms. Anchorage bad; the bottom sand. It has a small, deep lagoon at its head, connected by a shallow channel, which it has been proposed to deepen by dredging, so as to make an inner harbor completely shut in. The strong objection to this is, that it would become a nest for the yellow fever, which never leaves places thus shut in, and where there is no downward current, or heavy tidal ebb and flow to wash out the malaria. The country about Porto Caballos is fertile, well-watered, and wooded in the upland, but a dense, matted jungle along the coast, and exceedingly unhealthy. The prevailing winds are N. N. W., North, and West.

The highest elevation across this route is 2,016 feet above the sea. The grades of the proposed road are stated not to exceed 85 feet to the mile. The average temperature is 74° Fah.; annual variation, 68° to 88°. Distance from the Atlantic to the Pacific, 161 miles overland. In the interior the climate is generally dry and healthy.

The mineral productions are gold, copper, iron, and zinc. There is also white marble and limestone. The vegetable productions are sarsaparilla, vanilla, cocoa, pimento, sugar-cane, and indigo. The forests abound in mahogany, rosewood, lignum-vitæ, india-rubber tree, pine, oak, and various kinds of dye-woods. On the plains and in the valleys are tropical fruits.

The Southern or Pacific terminus of this route is the Gulf

of Fonseca, Lat. 13° 21′ N., and Long. 87° 35′ W. from Greenwich. This gulf has a number of excellent harbors, shut in and well-sheltered. It is about 50 miles in length, by an average of 30 miles in width. It contains four large islands, having ports with ample depth of water. Port "La Union" is formed by a small bay, and is the principal port of San Salvador. Amapala, on the island of Tigre, is the port of Honduras. Either of these ports has sufficient depth of water for vessels of the first class. The anchorage is in from 5 to 16 fathoms.

The valleys and plains of the Pacific side are subject to overflows, and the flooding was so great that the survey could not be regularly made by the English Commission, yet there seems no doubt of the feasibility of a road transit over the route; but from the vast filling in and embankments to avoid the overflow, it would be of enormous cost; so great, indeed, that the scheme has been for the present abandoned in England, where a company was formed for its construction.

3. THE NICARAGUA ROUTE.

Greytown, or San Juan de Nicaragua, is the Atlantic terminus, Lat. 10° 56′ 33″ N., Long. 83° 43′ 40″ W. from Greenwich. This port is open from W. to N. W. Formerly fairly sheltered, it has undergone destructive changes, and the washing away of the sandy point which formed part of its harbor has so filled up the entrance, that, at last accounts, there was but a depth of *four and a half feet* over the bar from the sea. Thus the harbor is practically closed, and the route destroyed.*

The river San Juan empties into this port over an inner

* A recent attempt has been made to reopen this route, but at this date, (October, 1863,) the M. O. Roberts line of steamers, which were to be engaged in it, have been publicly withdrawn—only *four* feet of water having, upon examination, been found upon the bar—the depth had decreased six inches since the examination of 1861; thus sustaining the theory that ultimately the depth upon the outer and inner bar will become equal.

bar with only three feet water at low tide. It has been proposed to unite the river San Juan and the river San Pedro, in the supposition that the union of the two would throw out a sufficient volume of water to remove the bar. It is a question whether the direct opposite would not be the result; for it is a known fact that, at the mouth of all rivers, bars are formed by the check of their currents from contact with the denser waters of the ocean. The size of these bars is generally in proportion to the volume of water moving through the river's channel.

The length of the San Juan River, with all its windings from Greytown upwards, is 96 miles, six and a half miles of which is obstructed by four rapids, caused by ledges of rocks stretching across the entire river.

The distance between Lake Nicaragua and San Juan del Sur is 17 miles. The country is mountainous, but the mountains are not high, although more difficult and worse for transit, because of their thin and sharp spurs, than if they were of high and steady elevation.

San Juan del Sur—in Lat. 11° 15′ 57″ N., Long. 85° 52′ 36″ W. from Greenwich—the Pacific port of this route, is small in dimensions, and open from the S., S. W., and W. The prevailing winds from June to October flow from the S. and S. E. During these months the access is not only difficult, but dangerous. The climate is humid and unhealthy. There are many stagnant lagoons and marshes along the margin of the lake and river, causing intermittent fevers, which, when contracted by persons from temperate regions, produce, when not fatal, great mental and physical debility.

The vegetable productions are similar to those of other sections of the great American Isthmus, but the country is less prolific than any other, and offers no inducements whatever along the line of the route to colonists even of Isthmian origin, and hence it is that it has never been settled.

Next in order is the projected

CHIRIQUI ROUTE.

The Northern terminus of this route is in Admiral's Bay of the Chiriqui Lagoon, in lat. about 9° 7′ N. and long. 82° W. from Greenwich.

The Chiriqui Lagoon is the grandest harbor of the world. It consists, indeed, of many harbors embraced in one, of vast dimensions. This has three entrances, clear, bold, and well defined; but as Commodore Engle, U. S. N., who was sent out specially to examine this harbor, has given an official report, it is perhaps better to refer to that. He says—at page 2, Exec. Doc. No. 41, 2d Sess. 36th Congress:

"Harbors incomparable, of a size unequaled, depth of water in their three main entrances for our largest ships to enter without pilots, and when fairly within can run on any course with certainty, and anchor in good holding ground, without fear of any one of the numerous injuries connected with the life of a seaman.

"The shores in most places are bold; should it be necessary to wharf for the accommodation of merchant ships, it will only be to give a solid, clean, and clear place to load and discharge, for they can now haul alongside the banks and secure to them. Wood for wharfing or other purposes is at hand. This also may be said of the Pacific harbor. The entrance to Golfito is perfectly clear, and the harbor for safety, or in points to facilitate trading vessels, quite equal to that of Chiriqui, but in size or grandeur neither that nor any other can be compared to it.

"The geologist satisfactorily shows the quantity of coal on the islands, but particularly on the tributaries of the Changuinola River, to be abundant, as well as of an excellent quality for commercial purposes.

"The nearest point from whence to ship the coal, I think, will be found to be Boca del Drago, which is the main channel to the beautiful, great, and grand Almirante Bay, thirteen miles long and eight broad, connecting with those of

Boca del Toro, Shepherd's Harbor, and Poros and Palos Lagoons. The two latter are more like large docks; their entrances are narrow, not over one hundred feet wide, with a channel of from eight to ten fathoms of water. When a ship is in her length, on looking around you discover yourself in a land-locked basin of three and a half by one and a half miles. Then at the foot of Almirante Bay, 'so called from its having been discovered by Columbus in October, 1503, where he remained ten days,' lies Shepherd's Island, cutting off one and a quarter miles of it by three and a half long, forming a complete place for the terminus of a railroad. There is a channel at each end of the island, giving to sailing vessels a fair wind at all times.

"This harbor, for its facilities for wooding, watering, and easy access, with its natural beauties, makes its mark on the minds of navigators never to be forgotten. The island is two hundred and sixty-four feet high, and can be made a spot of immense strength and beauty. The batteries from it would command the entrances as well as the whole harbor and terminus of a railroad on the southern shore. The average height of the highest points of the islands within the harbors is four hundred and fifty feet.

"Whenever this railroad is opened, of the feasibility of which the topographical engineer entertains no doubt, these harbors cannot longer remain idle, for the road will pass through a country of gold, coal, minerals, lands which produce every fruit of the tropics of the highest order—coffee, cocoa, indigo, rice, cotton, oranges, lemons, limes, &c. It is much easier to say what it does not produce than what it does, for everything seems to vegetate and give the best of its kind. Then, on the Pacific, the plains are large, and no bounds to the scope of pasture for raising cattle.

"At this early day I call the attention of the engineers and managers who may locate the road, to an examination of the line from the 'Three Branches' of the Robalo River to Shepherd's Harbor. The present proposed line will touch the 'Three Branches.' It is not much further, if any, from them to Shepherd's Harbor than to Frenchman's Creek.

"This harbor is near the coal-mines on the Changuinola River, where the valleys, by the report of the geologist, are thirty miles long by twenty broad, and where, on the Atlantic side, must be the position for active trade in coal, agriculture, cattle raising, &c. The hills are not so high, nor are they so rough, as those about Frenchman's Creek; some of them are now cultivated, and all can be. But the strongest recommendation is, that after entering Boca del Drago you have but twelve miles to run, passing through the most beautiful harbor of this magnificent group—no shoals, all clear.

"To Frenchman's Creek from Boca del Drago is forty-two miles. You pass through the Boca del Tigre, a fine, wide, open channel, which only requires a fort on Watercay to command it, and a light for navigators to clear the Tiger breaker, as well as for a point to determine the course to be steered through the shoals to the southward of it. Then you have twenty miles of smooth harbor surface to steam before reaching Frenchman's Creek, the now proposed terminus of the road, which would increase the steaming distance from any of our ports, *via* Cape San Antonio, thirty miles each way. Examine the line for Shepherd's Harbor. Boca del Drago has Columbus Island on its east and the main on its west. On entering the channel you make a sharp angular turn and head for the main, where there should be a fort; from this point you have two raking fires, which no ship or ships can stand, for you have to face the muzzle to a distance of fifty yards, then broadside, and then 'stern to.' Near this point must be the great depot for the coal from the mines, which must be brought there for shipment. It is the most available point; the distance to Changuinola is six miles."

There is now no doubt whatever of the feasibility of a railroad across from the harbors of the Chiriqui Lagoon to the Pacific harbor of Golfo Dulce. Lieut. Morton, topographical engineer, U. S. A., under order of Congress, made a reconnoissance of the route in 1860, and his report is set forth in Exec. Doc. No. 41, 2d Sess. 36th Congress.

Lieut. Morton's observations were directed to a single pass—a cañon then recently discovered, and he went through only ONE branch of this—and that, as it is asserted by residents who have since gone over, the most unfavorable of all. Beside these cañon routes, routes which cut down *both the mountain and the upper plains* in their altitude, there are others quite as favorable, if not more so, than these deep "baranca" or cañon passes. The country is, however, a wilderness, almost unknown, and, until these Chiriqui grants of lands were made, had not been surveyed, or even crossed except by the native Indians.

In 1852 James B. Cook, an English engineer of reputed ability, was sent out by a Company formed in London, to examine this country, and to ascertain if a carriage road was practicable across the Isthmus of Chiriqui. A copy of his report may be found in full at page 46, Report No. 568, (U. S.) House Representatives, 36th Congress, 1st Session. Mr. Cook surveyed with a theodolite, compass and chain. His highest grades through the mountains are "*one in fifty*," which is 105 feet to the mile. This is *seven feet lower* than the *most favorable* MOUNTAIN grades of the Baltimore and Ohio Railroad. The other mountain grades are "one in sixty" and "one in sixty-five feet." These are, respectively, 81 and 88 feet to the mile, and these grades extend over only about 14 miles of the road. The remaining grades, he says, will not exceed 20 feet to the mile, while even the higher ones can be reduced by cuttings and embankments.

Professor Manross was sent out in 1854 by the Chiriqui Improvement Company. His ability was vouched for by Professors Silliman and Hitchcock, as a geologist. His duties were to examine the coal formations. After accomplishing these he crossed the Isthmus to the Pacific, made observations with the barometer upon the altitude of the plains and the summit pass. His report is given in full at page 40, House Rep., Report No. 568, First Session 36th Congress. He says the ascent is regular, not exceeding three feet in the hundred, quite up to the summit. This would give $158\frac{4}{10}$ feet to the mile as a barometrical measurement of the highest point, and allowing for the usual deflections and curves by which a road is lengthened to secure the best grades, corresponds almost exactly with Cook's statement.

Lieut. Morton says the highest grade over *his* line is 300 feet; *that this is a temporary one only* to cross the *mountain summit ridge,* while a tunnel is being cut through the mountain below, *which tunnel will take off* 2,200 feet of the elevation, and reduce the maximum grade, at its sharpest point of ascent, greatly below 200 *feet* on a SURFACE grade, without allowing for cuttings and embankments; and he says that the "Garuma" route is about 700 *feet lower still* in altitude, but not quite so favorable to connect with the Pacific plains. Mr. Jeckyll, the surveyor who accompanied Mr. Morton, and who appears to have done most of the work, says that Morton's reported grades can be reduced from 20 to 50 per cent. (see his letter to Hon. F. H. Morse, Chairman Committee on Naval Affairs.) This statement, therefore, of Mr. Jeckyll's corresponds almost exactly with those of Cook and Manross.

There is, then, the concurrent testimony of three able and scientific men, unknown to each other, each acting at a different period of time, (the personal and scientific character of each well vouched for by parties of high position and wholly disinterested,) that the route is entirely practicable for a railroad with grades no higher than one of the hitherto most successful freight and passenger roads of this country. To these are to be added the express declaration of Lieut. Morton, the U. S. Government Engineer, who says, at page 42, part 3d, Message and Documents of President of U. S., 1860 and 1861,—"I feel warranted, however, in "stating at this time, as the result of my examination, the "conviction that *it is entirely practicable to connect the har- "bors by a line of railroad* ADAPTED TO COMMERCIAL PUR- "POSES. The savannahs which stretch from the Pacific "coast to the summit of the line offer a bed for the track, and "a natural grade that needs but little improvement. Both the "plains and mountains abound in timber suitable for framing or cross-ties."

Dr. Evans, the Government Geologist, in speaking of the coal, *the road*, and the harbors, says:

"Deposited, as this extensive and almost inexhaustible coal-field is, at the largest and safest harbor on the Atlantic coast, opposite to another harbor on the Pacific capable of receiving at safe anchorage all the shipping engaged in commerce between Europe and Asia; with a country rich in mineral resources; tropical fruits in abundance; with every variety of soil and climate as you ascend the mountain ranges; abundance of turtle and fish in the waters adjacent; plenty of game on the mainland, the islands in the lagoon, and on the Pacific coast; with three practicable railroad routes across the Cordilleras, in addition to the cañon route surveyed by Lieutenant Morton, as I can state from personal observation, affording the best connection for the commerce of the two oceans, this country offers a wide field for American enterprise, and is well worthy of the patronage of the Government."

(See Exec. Doc. No. 41, Second Session 36th Congress, page 46.)

This latter statement is authenticated by Commodore Engle, of the U. S. N., and Chief of the surveying expedition, who says, at page 4, same document:

"If there had been sufficient time, Lieutenant Morton would have been directed to examine and survey the other passes designated by Dr. Evans, which, he states, were discovered or observed by him during his geological explorations; and that they have lower grades than the one surveyed by Lieutenant Morton. The twelve years' experience which Dr. Evans has had as a Government geologist and surveyor in the Rocky Mountains entitles his opinions to consideration.

" The results of the explorations and surveys made by the Chiriqui commission are thus demonstrated: by Lieutenant Morton, that a practicable route for a line of railroad has been found between the two oceans, on the Isthmus of Chiriqui.

" By Lieutenant Jeffers, the hydrographer, that the great and grand harbors at its termini afford every requisite for the protection of naval and commercial marine, and for all practical purposes, to an unlimited extent. And by Dr. Evans, the geologist, that the best coal for steam navigation exists at and near the Atlantic harbors of the Chiriqui Lagoon."

In all the records of railroad making, it would be difficult to find stronger concurrent testimony than is thus given of the practicability of this route; testimony, too, by parties unknown to each other, and at intervals of time running over several years.

There is no other portion of the Great Isthmian strip, or of Central America, so healthful as Chiriqui.

Mr. Cook, the English Engineer, ascribes this to its proper causes—the lofty mountains, elevated plains, and the absence of marshes. In his report, (see page 47,) House Naval Committee Report, No. 568, 36th Congress), he says:

" Throughout the whole extent of the road there is a succession of mountain streams, none, however, presenting serious obstacles to its construction, but affording to the road a supply rarely found in elevated countries.

" Though rapid in their courses, these streams do not run out or exhaust themselves in the dry season. The reason of this is obvious; the peaks of the lofty mountains, rising some thousand of feet above the line of the road, penetrate the dense masses of vapors thrown up by the rapid evaporation of tropical climates, and form conductors around which is condensed the irrigated supply of the valley below. To this constant coursing of water may also be attributed the extraordinary healthfulness of this entire route across the isthmus; no other compares with it. The crossings of Tehuantepec, Guatemala, Nicaragua, Darien, and Panama, are, throughout nearly their whole extent, subject to fevers and maladies of the most fatal character; the decaying matter of their swamps and marshes impregnate the air with their destroying miasma, and the adventurous stranger who seeks the alluring gold-fields of California and Australia, too often finds his grave upon the narrow strip which divides these great oceans. Not a marsh, or swamp, or putrefying place exists from ocean to ocean on the Isthmus of Chiriqui. The Chiriqui road will then, when finished, from these natural advantages, secure to itself the great travel, not only from the cities of the United States to California, but also from England to Australia and the East Indies. Nor does the province of Chiriqui offer less inducements to the emigrant to settle within its borders than will its road to the voyagers to distant lands."

"Rich and fertile plains, from which spring, in the most luxuriant growth, the coffee-tree, yielding in abundance, after three or four years' culture, a berry of extraordinary flavor. The tea-plant growing wild, and requiring but the hand of industry to make it most productive; Cotton—rice of large grain, *tobacco*, flax, maize or Indian corn—the English or small-grained corn; wheat, Irish and sweet potatoes, peas, beans, cabbages. Besides these, there are various spices; including ginger, pepper, and a great variety of botanical and medical plants. The slopes of the mountain ranges afford pasturage of the best description, and thousands of cattle feed wild upon their sides."

Dr. Evans, the U. S. geologist, at page 53 Exec. Doc. 41, says:

"There are no prevailing diseases between the Atlantic and Pacific coasts at the Isthmus of Chiriqui. During the whole of my explorations of the shore line of the Lagoon—the islands adjacent—the various rivers tributary, and in crossing and recrossing from the Atlantic and Pacific, not a single member of my party, either the men from the 'Brooklyn,' or natives, was sick. It is true, that at the unfavorable locality of the 'Mission House' at the mouth of Fish Creek, with a marsh back of the settlement filled with water at every rain, and covered with vegetable matter in a state of decomposition, washed down from the adjacent mountains, constantly accumulating, and subjected to the heat of a tropical sun, cases of intermittent fever occurred, but they readily yielded to a few doses of quinine. Fatal cases, if any occur, are rare. A similar unfavorable locality in any section of country would produce as great, if not greater, deleterious effects. My own hammock was swung, for ten nights under a palm-leaf roof, elevated ten feet from the ground, with the land breeze passing over this swamp, and I suffered no inconvenience, and not a moment's ill health.

"The rainy season is a succession of sunshine and showers. If it rains in the morning, it is usually fair in the afternoon. And if it rains in the afternoon and night, it is bright in the morning and until three or four o'clock in the afternoon. Sea breezes prevail during the day, and land breezes at night. The thermometer varies from 67° to 87° during the year, and towards morning it is so cool, that a blanket was comfortable every night during my sojourn on this Isthmus.

"In describing the formation of the soil, I have already briefly alluded to the productions of the country; but it will not be out of place to refer to special localities and their rich producing qualities.

"On the Atlantic side, there are many streams flowing down from the mountains having bordering valleys; but the most beautiful and extensive are those of the Cricamola and of the Changuinola rivers. These vary from ten to twenty miles in width; their soil is inexhaustible. Indigenous to it are cotton, tobacco, coffee, cacao, sugar-cane, rice, and all the tropical fruits. These do not grow in single crops, but in succession throughout the year. Under the hand of the cultivator, four crops a year could readily be gathered, and of qualities inferior to none that earth produces."

The Pacific terminus of the route is *Golfito*, a harbor opening out from Golfo Dolce. This has already been alluded to in the extract from Commodore Engle's report, but the U. S. hydrographer, Lieut. Jeffers, thus describes it (at page 43, Exec. Doc., No. 14):—"This beautiful harbor, situated twenty miles from Punta del Banco, on the eastern side, and midway between the entrance and bottom of the Gulf of Dolce, is unsurpassed in natural facilities. Favorably situated for entering with the sea breezes and leaving with those from the land, both of which are regular. There is no bar or other obstruction at the entrance, which is upwards of half a mile (1,200 yards) wide, and about a mile in depth, with an excellent anchorage in good holding ground, in five, seven and twelve fathoms; having the chart, no other guide than the eye is necessary. This outer harbor is separated by a sand spit, a mile in length, by a few feet in width, (around the northern extremity of which, there is an excellent channel eight hundred yards wide, with upwards of five fathoms of water,) from the inner harbor four miles in length, with an average breadth of one mile.

"The inner harbor has about a square mile of anchorage, with a depth of five fathoms, (sufficient for the largest ships,) and about three square miles of anchorage for vessels of smaller size.

"On the northeast side of the harbor, opposite the entrance, there is a range of hills several miles in length, parallel to the coast line, of an average elevation of about fifteen hundred feet, leaving a strip of level, generally but a few yards wide, between their bases and the shore line. On this side of the harbor, and for a distance of upwards of three miles, not less than five fathoms of water is found within half a cable's length of shore, affording ample room for wharves sufficient to accommodate any probable number of vessels frequenting the harbor, if the proposed road should be built.

"Three streams—the Golfito at the eastern, and the Corisal and Cañaza at the northwestern extremity—empty into the harbor; neither of these navigable, but either affording an ample supply of fresh water for the proposed town.

"There is one level spot sufficient for the site of a large town; but the several valleys running back from the shore in the aggregate would give about a square mile of building sites. The whole surface, to the tops of the highest hills, is densely wooded, excepting the sand-spit, and a small spot of about three acres, near the entrance, under cultivation. A great variety of these woods are valuable for ship-building or for other constructions; different woods excelling in the qualities of durability, hardness, and strength."

The temperature is represented as cool and agreeable: average 6 A. M., 74°; meridian, 83°; 3 P. M., 84°; 6 P. M., 78°.

This harbor was also surveyed under orders of the Emperor Napoleon, by Captain Colombel, September, 1851, and that survey verified in June, 1852, by the French Admiral Odet Pellion; they unite in closing a very full description of it in this expression: "The Golfito is an immense natural basin, or great bay of deep water, and would make the finest military post in the world." See House Rep. Report, 1st Sess. 36th Congress, No. 568, pages 74–76.

THE PANAMA ROUTE.

Aspinwall, the Atlantic terminus of this route, is situated at the upper part of Navy Bay. This bay is formed by an irregular semicircular indentation of the coast, southward from its general lines; the broadest portion of the bay is its entrance, and that faces nearly due north; consequently it is completely unprotected from the prevailing winds, which are from the north, and even a moderate sailing breeze throws up a heavy sea upon the structures intended for wharves, so that they cannot, save in dead calms, be used for landing purposes. During a gale, or "norther," no vessels can lie in the port with safety, but are forced to weigh anchor, not unfrequently to slip their cables, and proceed to sea. Porto Bello is their nearest place of refuge. Wrecks are not unfrequent within the very harbor of Aspinwall, and it is entirely out of the question to attempt a landing of freight or passengers during the slightest northers. The town of Aspinwall is located upon a small coral island, known as "Manzanilla," which is some five or six hundred feet from the main shore, and is connected therewith by the railroad bridge. The place has no fresh water, and the inhabitants, as well as the vessels to the place, are dependent upon and forced to send to Porto Bello, some twenty-five miles distant. And even this supply is a monopoly, the water being brought down in pipes to the shore, and sold at a high rate by the Steamship Company, its alleged owners.

The Panama Railroad extends in a southeasterly direction across the Isthmus, its length between the harbors, Navy Bay, and Bay of Panama, being forty-seven miles. The long, patient, and expensive effort devoted to the construction of this road, is worthy of all praise. Its construction is conclusive proof that there is no physical obstacle on land which may not be overcome by modern science and energy.

The line of country traversed by it is almost a continuous swamp, and where the swamp is intermitted and higher lands occur, especially hills—these hills being of decomposed porphyry—the cuttings and embankments of the road were subjected, and still are, to slides and crumblings, while the portions extending over the swamps have never had stability. A gradual settling down of the structure exposes it to frequent inundations; and recently, within a few months, a portion of the road was rendered impassable from this cause. These evils cannot be effectually remedied; and hence, this road would never answer for transporting large masses of heavy freight, even if the harbors were suitable for its reception and delivery. It will meet the requirements of a passenger and light freight route, until some great disaster, such as the sudden sinking of a portion of the road, with a passenger train thereon, will drive even this business to a surer route. Such a disaster may not occur; but its occurrence is neither impossible nor improbable, and there have been serious indications of such an event.

The climate along this route is deadly. The miasmatic effect, known as the Chagres fever, extends from ocean to ocean, and it is almost fatal to any one from a Northern climate to remain even a few days at any one point. If he escapes with life, he carries away a broken constitution, with the certainty of a return of the fever upon any slight exposure.

The City of Panama, the southern terminus of the route, is situated upon the bay of the same name, both too well known to require any lengthened description. The bay affords but slight harborage; large vessels cannot approach nearer than five or six miles to the shore. All reception and discharge of freight is performed by lighters—and these can only be employed at high water; for the ebbing tide lays bare three miles of foul, offensive, muddy bottom, which, at that time, renders approach from the sea to the city impossible. A wharf has been constructed out a portion of the distance, and it was contemplated to extend a mole out to deep water. Engineers, however, estimated that this work would equal the original cost of the entire railroad, and would even then be uncertain as to durability; hence it was abandoned, and the road exists without an available harbor for a great commerce at either terminus.

Two other routes have been proposed. The Darien and San Miguel, and the Atrate and San Juan. Expensive and disastrous surveys of these have produced only such information as to cause their entire abandonment.

The problem of a "Shorter route to the Indies," a direct means of communicating from the Atlantic to the Pacific, has from the earliest date of commercial enterprise engaged the earnest attention of nations, and of individuals. To solve the problem, Columbus steered his western course, thus discovering the continent we now inhabit; and it is a most remarkable fact, that the first portion of the main-land touched by him was the true point for its solution. This was the Isthmus of Chiriqui—the harbor he entered the "Almirante Bay," in the Chiriqui Lagoon.

Like other nations, our own has given careful attention to this problem, and we find that Mr. Jefferson thus wrote: "The report of a survey of the Isthmus of Panama is to me a vast desideratum, for reasons political and philosophical;" and Mr. Clay, when Secretary of State in 1825, wrote:

"The deepest interest is taken by the United States in the execution of an undertaking (the Nicaragua Canal) which is so highly calculated to diffuse an extensive influence on the affairs of mankind."

Mr. Forsyth, Secretary of State in 1835, says: "Mr. Biddle, whom President Van Buren has thought proper to employ specially for the purpose of gathering information relative to the prospects for connecting the Atlantic and Pacific Oceans;" and President Taylor, in his message of 1849, says:

"A contract having been concluded with the State of Nicaragua, by a company composed of American citizens, for the purpose of constructing a ship canal through the territory of that State, I have directed the negotiation of a treaty with Nicaragua, pledging both governments to protect those who shall engage in and perfect the work. The routes across Tehuantepec and Panama are also the subject of our serious consideration."

From the year 1849 to the present time, surveys and experimental openings of supposed good routes have followed each other in rapid succession, until the information set forth in the foregoing description of each has established the following results:

1. That the geographical position of Tehuantepec would afford many advantages for a postal and military route from the Southern ports of the United States; but the radical defect in its harbors, and the impossibility of improving them, deprive it of all consideration either for these uses, or for the commerce of the North Pacific. If the harbors were good, other and lower routes would have the entire advantages in competing for the South American and the South Pacific trade.

2. The Honduras route, extending through a section of country almost entirely under the control of Great Britain, even were it positively practicable, would be of little advantage to the United States; but the narrow, confined and extremely unhealthy harbor at its Northern or Atlantic terminus, together with the vast cost of constructing the road, has caused its present consideration and construction to be abandoned by its British projectors.

3. The Nicaragua route, under its many changes, and the zealous efforts of each of its possessors to extend it to commercial capability, has failed. The quicksand formation of a large portion of the San Juan River, and the rocky ledges which in other places break it into falls and rapids, defy practical improvement. The darling scheme of a French Emperor, the fostered enterprise of American energy, it has baffled each, and finally, the utter destruction of its Northern harbor by the surging waves of the Atlantic, has removed it from competition with other routes; and until the currents of the river and the waves of the ocean shall reunite

in forming a harbor at San Juan or its neighborhood, "the Nicaragua route" is of the *past*.

The Panama route is now the only existing transit across the Isthmus. Notwithstanding the great and irremediable defects in its harbors, it has answered the great purpose of mail and passenger transit, and thus far a safe means of transmitting the gold of California to Atlantic ports. The almost barbarous condition of the population of Panama offers, however, but slight guarantees for the future; and while the mob spirit of the place threatens life and property, the defective harbors afford but little means for the United States to protect its people and treasure, while in transitu; and the hazards have increased with our political difficulties. A further important consideration is, that this route, in twelve years, will pass to the control of Great Britain. New Granada reserved in the grant the right to take the road at the end of twenty years, by paying the grantees for it (and its equipments, all of which were to be kept in complete order) the sum of five millions of dollars. The road and property cost ten millions. The necessities of New Granada forced her to borrow money in London upon this right of resumption, or foreclosure, which was pledged as the security. Without revenue, that government cannot repay this debt; and the result must be, that the parties, in 1874, will take possession of this inter-oceanic highway, under the name of a British company, paying to the present Panama Company the amount stipulated in the grant to them. The United States can interpose no obstacle to this—it is a legal right reserved by New Granada, and transferred by her to British subjects.

Vast interests to the United States are dependent upon an inter-oceanic transit, held exclusively by her people, over the American Isthmus. Whether this be at its sections—the Isthmus of Tehuantepec, the Isthmus of Nicaragua, the Isthmus of Panama, or the Isthmus of Chiriqui—is wholly immaterial, so that the *conditions* of transit route are obtained. The conditions or qualities are, good harbors at the termini, a solid foundation for the road, a healthful climate, and facilities to protect the people and property to be passed over it, or while at its termini.

Motley, the historian, in his late letter upon the American troubles, says: "The division of the national domain, the " navigation and police of the great rivers, the arrangement " and fortification of the frontiers, the transit of the Isthmus, " the mouths of the Mississippi, the control of the Gulf of " Mexico—these are significant phrases, which have an ap- " palling sound, for there is not one of them which does not " contain the seeds of war." Mr. Motley might have added to his sentence: "Seeds which can only germinate through the weakness of the Nationnl Government in failing to place each of the three first in its relative position of strength, and in permanently securing and guarding the three last by possessing, through its citizens, the proper and effectual point for naval strength and inter-oceanic transit upon the Isthmus."

The 36th Congress, by authorizing a commission to examine and report upon what was believed to be the best Isthmus crossing, has placed the true information as to its merits before the Government. That commission, composed of distinguished officers of the navy, army, and civil service, in their report, leave no further room for doubt. That report shows, that

The Isthmus of Chiriqui has "harbors incomparable, of a " size unequaled, depth of water for the largest ships to " enter without pilots; when fairly in, to run on any course, " and anchor without fear of *any one* of the numerous inju- " ries connected with the life of a seaman."

A route between, and connecting these harbors, which, though difficult in parts of its length, is yet feasible throughout the whole; and the road, when made, will have a solid and durable foundation, and this through a country healthful, rich, productive, all that can be expected or desired in the luxurious abundance of a tropical region.

The legal rights of property in these lands, harbors, route, &c., is in the hands of American citizens.

These *harbors*, the *coal*, the *healthful road route*, stamp this as the highway of nations between Europe and Asia, between the United States, Asia, and South America. If this highway is controlled by the people of this country, it will largely control the commerce of all countries. Cities will be founded and grow up at its harbors, and from them will radiate an influence which cannot now be appreciated.

The harbors and route are beyond the limits of the Clayton-Bulwer treaty, and free from its entanglements. Either the United States, or Great Britain acting under the grants of the Chiriqui Improvement Company, may control the enterprise and found these cities. Whichever shall do so will thenceforth hold the scales of commerce in her hands, to mete out such portions as she may determine to other nations.

The Chiriqui route has the practical advantage in time or in distance over all others. This will be apparent by drawing an accurate scale of degrees upon any globe. It will be seen that the distance is—

From Liverpool to San Francisco, via Panama............................7,980 miles.
From Liverpool to San Francisco, via Nicaragua............................7,720 miles.
From Liverpool to San Francisco, via Tehuantepec............................7,440 miles.
From Liverpool to San Francisco, via Chiriqui............................7,390 miles.

In every route but Chiriqui the sailing distance is increased by the deflection of course round the West India Islands, or the projection of Santa Mala, which forms the Bay of Panama, in the Pacific. To and from the harbors of Chiriqui, on the Atlantic and Pacific, are direct lines, without bend or deflection.

If the Tehuantepec were *a practicable* route, then it would be shorter in distance via New Orleans, (the inland route thereto,) but as a sea route, with its deflection round Florida, and upwards into the Gulf, it is rendered of nearly the same length; while its entire want of harbors and coal for steamers places it at once outside of any estimate for general commercial purposes. The actual distances from New York to San Francisco are as follows:

New York, via	Panama,	to San Francisco..	5,224	miles.
"	Nicaragua	"	4,700	"
"	Tehuantepec	"	4,200	"
"	Chiriqui	"	4,684	"

From Europe to Australia or China the Chiriqui route has an average advantage of about three hundred miles shorter distance than *any other*.

When the actual saving of distance, of time in entering and in delivering of passengers and cargoes without the necessity of resorting to boats or lighters, and the *local supply of coal* at rates to defy competition in the other places named, together with the healthful climate, are taken into consideration, it will at once be apparent that the Chiriqui stands without a rival.

www.ingramcontent.com/pod-product-compliance
Lightning Source LLC
LaVergne TN
LVHW011142110826
845150LV00008B/2462
* 9 7 8 1 4 1 8 1 9 3 5 2 2 *